A LITTLE ROCKET SCIENCE
AND OTHER BAUBLES

by

Olatunde Adeyemo

©2013

Lulu.com 2013

© Olatunde Adeyemo 2012, 2013
All rights reserved, No parts of this publication may reproduced, or stored in any retrieval system, or transmitted in any form or means, electronic, mechanical, photocopying, recording or otherwise without the prior written permission of the author.

First published 2013

ISBN 978-1-291-37481-0

Abstract

A LITTLE ROCKET SCIENCE AND OTHER BAUBLES

Olatunde Adeyemo BSc MSc CIWEM CEng.

In pipe network analysis we seek to find a steady state system of flows that will allow delivery of water to where it is required in quantities sufficient to meet local demands. In a single loop problem, we discover it is just a question of pumping the fluid with enough pressure to overcome the frictional headloss due to momentum forces, wall roughness and the requirement that the fluid must slide over the the pipe internal surface which offers some resistance to the flow. There are two formulae in general use for calculating headloss in sealed pipe environments, the Darcy Weisbach and Hardy Cross formulae. I use both to show the relative ease of using them to determine the flow through a single loop pipe network when given a specific pumping pressure. I demonstrate how the problem of solving the flows in a multi loop network can be solved using the same method, by analysing the network into loops and iteratively solving a single loop problem on the worst matched loop.

This allows for a simplification of computation making it unnecessary to have to solve simultaneous equations and obtaining an accelerated convergence to the solution. It also means that the formulae relating the headloss and flow do not have to be inherently linear and we are able to get solutions for the complex Darcy Wiesbach formula which should give more realistic results. The main point of this method is it relies on the analysis of the network into loops and creation and setting out of initial conditions.

If we have a system of nodes and connectors such as a pipe network or a traffic system or even a structural lattice, as long as we can define mathematically how properties are transferred between nodes and the capacity of the connectors we can use a heuristic approach to solve the network and produce a steady state equilibrium solution. Comparing such a solution with real time monitoring to find how far a system is away from equilibrium will indicate the degree of stability, where we can expect bottlenecks and congestion and the forcing mechanisms within the system. In this volume we demonstrate how this type of solution is is used to look at a lattice of elastic spars. We get a graphical solution to the equilibrium position of the lattice loaded under different conditions.

In the last section of this volume we look at the classical rocket problem. A small payload sited on a massive fuel package. How much fuel do we need to put an object in orbit? We derive estimates for energy requirements for different orbits and show how different fuels can have implication on the different rocket sizes.

Table of Contents

List of figures..iv
Acknowledgements...vi
Pipe Network Analysis From First Principles.......................1
    1.0   The problem..1
    1.1   Networks...2
    1.2   Single Loop Problem...2
    1.3 Multiple Loop Networks..11
A Heuristic Approach to solution of Equilibrium in Statically Indeterminate Structures...25
A Little Rocket Science..42
    3.1 Rocket Momentum Consideration..........................42
    3.2 Energy Conservation..45
    3.3 Energy Considerations...50

Number   Page

Table of Figures

Figure 1.21   Single Loop..3
Figure 1.22   ΔP over a length of Pipe............................4
Figure 1.31   Example Network..................................14
Figure 1.32   Loops and flows....................................17
Figure 2.1 Bending Moment..28
Figure 2.2 Unstrained Lattice.....................................29
Figure 2.3 Load =-10 N...36
Figure 2.4 nodes 2,3 free...37
Figure 2.5 force =-25N 38
Figure 2.6 Force =-25N nodes 6,7,8 free....................40
Figure 3.1 Momentum Considerations........................43
Figure 3.2 Energy Conservation.................................45
Figure 3.3 Locus of Orbit..51
Figure 3.4 Diurnal Rotation..56

List of Tables

T 1.21   Solution of h=25m Darcy Wiesbach...............9
T 1.22 Solution of h=25m Hazen Williams................10
T1.31   Flowchart for analysing Elementary Networks
....12
T1.32 List of nodes with properties...........................15
T1.33   list the pipes and their properties,..................16
T1.34 Network Loops..17
T1.35   A Darcy Weisbach solution to Pipe Networks....18
T 1.36 Input File Content of NsLpIn.txt.......................21
T2.1 Steps in Forming Approach................................26
T 2.2 Unstrained Node Positions................................30
T 2.3 Spar properties...31
T 2.4 Program to determine equilibrium positions of free nodes..34
T 2.5 nodes with largest forces..................................39
T 2.6 Struts with largest forces..................................40
T3.21 Stoichiometric Isp 49
T3.31   Location of Sites...55

T3.32 Energy of Diurnal Rotation..............................56
T3.33   Escape velocities of celestial bodies................59
T3.34 Effective Specific Impulse..................................60
T3.35 Volume of fuel for 5000kg Payload...................62

ACKNOWLEDGEMENTS

I wish to acknowledge the great influence and inspiration instilled in me by my past teachers and tutors. They contributed to motivating and instigating me in my quest for scientific advancement and development. Notably among them are Miss Gamgee of Sellincourt Primary School, Mr D Price of Battersea Grammar School, Dr D Finlayson, Professor A Maitland, Professor J F Allen, Dr Bill Hanbury, Professor A Cracknell, Dr P Davies, Dr J Crowther, Mr W N Okutu, Engineer A O Omolukun.

Introduction

This volume is a continuation of my previous work "Matrix Magic", it continues to demonstrate practical uses for a mathematical approach to science with evaluated examples. At first we examine the solution of pipe networks, using linear Hardy Cross and non-linear Hazen Williams formulae.

We demonstrate how a heuristic system can be developed and is transferable to problems of different disciplines. We use it to solve the problems of statically indeterminate structures and show how it is important for the analysis of modern light motile structures that are to be found in the aerospace industry and support structures that work under fixed strains such as bridges and towers.

Finally we look at the subject this volume takes its title from, i.e. rockets and objects in orbit. We see how consideration of momentum and energy lead to implications of fuel usage and the feasibility of different orbits. We conclude by making minimum estimates for various fuels delivering a 5 tonne payload into a low earth orbit.

Chapter 1

PIPE NETWORK ANALYSIS FROM FIRST PRINCIPLES

1.0   The problem

We wish to use a pipe network analysis to examine a sealed system of pipes, reservoirs, pumps, sources and demands and find the steady state pipe flows and pressures, i.e. what pipe flows and pressure losses could be expected to be sustained under the conditions?

At first impression one might think the solution is easily achievable, surely all that is required is the correct algorithm between flow and headloss. The problem is that we deal with networks and there are more than one route between two points. For the solution to work, a single value of pressure difference is expected whatever route taken between the two points. The solution is going to be achieved by suggesting a solution that will satisfy the supply / demand requirements. The pressure losses through the system are then calculated and the variation between different routes are measured and a correction is then calculated and summed to the solution. The process is repeated until the discrepancies lie within tolerance. The details of the iteration process depend on the algorithms of head loss and method of iteration.

1.1 Networks

Network consists in the main of components pipes, connections, supplies and demands. Each pipe forms a connection between two nodes, nodes may be a supply or reservoir with a free surface pressure, or connection of pipes with intrinsic demand or supply.

Pipes have properties of length, diameter and some factor describing roughness that is dependant on the condition, method of fabrication, age and type of material that comprises the pipe and the method of pressure loss calculation. The pipes have to be the enumerated with their properties.

1.2 Single Loop Problem

For the simplest type network to illustrate methods of calculation we may examine a single loop problem. Consider the following single loop problem.

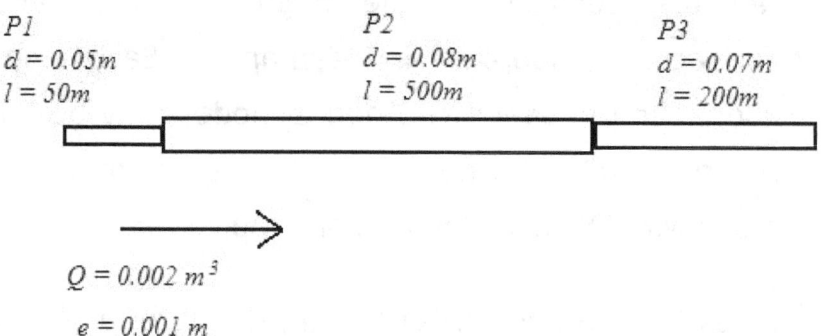

Figure 1.21   Single Loop

We can easily solve this series of pipes since the pressure drop in each pipe is calculated and the results are summed. We will consider two methods of calculating pressure drop or what we alternatively call headloss.

The D'Arcy Weisbach formula is a method that bases the calculation of headloss from the analysis of a level circular pipe with the variation in cross-sectional pressure due to the bulk fluid moving at average velocity and the friction due to shear stress at the pipe wall. This is expressed as

Eqn 1.21 $$\frac{\Delta P}{\rho} = \frac{flv^2}{2d}$$

Where f is an empirical dimensionless constant called the friction factor which is dependent on Reynolds number (R_e) and e/d from the Moody's diagram. When we consider specific energy (E/m) having dimensions (l^2t^{-2}) we see this is the same as P/ρ, thus the change in specific energy due to friction is (ΔP/ρ) this has the same dimensions as h * g, as such specific energy can be considered as "head of water".

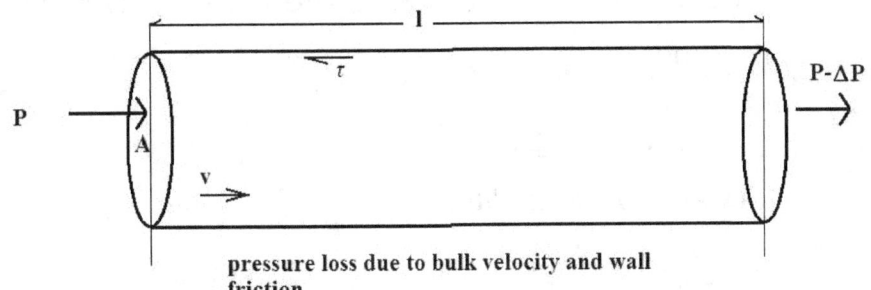

Figure 1.22 ΔP over a length of Pipe

Eqn 1.22 $\quad \dfrac{\Delta P}{\rho g} \equiv \Delta h$

Δh is headloss and

Eqn 1.22b $\quad \Delta h = \dfrac{flv^2}{2gd}$

Bernoullis equation for an incompressible fluid gives us

E1.23 $\quad P + \rho g z + \dfrac{\rho v^2}{2} = \text{constant}$

Thus we can assume across pipe face at l

Eqn 1.24 $\quad \dfrac{P - \Delta P}{\rho} + gz + \dfrac{v^2}{2} = \text{constant}$

To derive f we may use the graphical method using e/d and R_e and the Moody's chart. However with moderate computing ability we can find f as the solution to the function F=0 which is solved using Newton Raphson method

$$F(X)=0; \quad X = x+h;$$
$$F(x+h) = 0 \approx F(x)+hF'(x);$$
$$h \approx \frac{-F(x)}{F'(x)}$$

where

Eqn 1.25 $\quad F = 11.14 - \dfrac{2}{\ln 10} \ln\left(\dfrac{e}{d} + \dfrac{9.35}{R_e x^{1/2}}\right) - \dfrac{1}{x^{1/2}}$

And

Eqn 1.26 $\quad F' = \dfrac{1}{\ln 10} \dfrac{\dfrac{9.35}{R_e} x^{-3/2}}{\dfrac{e}{d} + \dfrac{9.35}{R_e x^{1/2}}} + \dfrac{1}{2} x^{-3/2}$

Applying this to our 3 pipes of the example.

Pipe 1

Q=0.002 , d=0.05 , e=0.001 , l=50

$$v = \frac{4Q}{\pi d^2} = 1.02 \text{ m/s}$$

$$R_e = \frac{\rho v d}{\mu} = 5.09 \times 10^4$$

f = 0.0493

$$\frac{\Delta P}{\rho l} = \frac{fv^2}{2d}$$

$\frac{\Delta P}{\rho l}$ = 0.513 , h=2.62

Pipe 2

Q=0.002 , d=0.08 , e=0.001 , l=500

v=0.398

R_e=3.18 x10^4

f =0.0425

$\frac{\Delta P}{\rho l}$ =0.042 , h=2.14

Pipe 3

Q=0.002 , d=0.07 , e=0.001 , l=200

v=0.520

R_e=3.64 x10^4

f =0.0442

$\frac{\Delta P}{\rho l}$ =0.0853 , h=1.74

Thus the total headloss associated with the pipe and flow arrangement is 6.50m. This does not include minor losses such as entrance and leaving losses and losses due to connections such as reducers.

If the same network is defined with total headloss across the three pipes and an unknown flow, our desired solution would be to calculate the flow that produces the correct the individual headloss across each pipe. Thus we are looking for

Eqn 1.27 $$F = \frac{8Q^2}{g\pi^2}\left(\sum_i \frac{f_i l_i}{d_i^5}\right) - h = 0$$

Because of the irrational and extended nature of Eqn 1.27 we would find it difficult to derive a direct equation for flow given headloss. The method we use to solve this is an iterative process, where we make an initial guess of the flow, calculate the headloss due to this flow and evaluate the difference between this and the required headloss. This difference is used to generate a correction which can give us a better flow estimate with a closer total headloss. As long as the initial estimate lies close enough to the true value, that is, if it lies within a zone of convergence, we should expect subsequent estimates to converge to a solution.

For example, suppose from the same network we require a flow that gives a total headloss of 25m. We know a flow of 0.002m³ gives a 6.50m headloss. We may make a second guess at the flow creating a second evaluation. The linear estimate of the zero created from these two estimates would normally give a better estimate. By iteratively choosing the two points with the lowest difference between calculated head and required head to make subsequent estimates for the zero normally leads to a rapid zooming in onto the solution flow. This type of method would normally work with a function as long as our initial guesses are within a radius of convergence for the same root.

Consider the headloss at 0.00202 m³. We can calculate this and we can use the difference in the two calculations to create a new estimate for Q. Remember the function f is irrational depends on Reynolds number which in turn is dependent on Q. We need to solve

$$\sum_i \frac{f_i l_i}{d_i^5},$$

$f_1 = 0.0493$, $d_1 = 5 \times 10^{-2}$ $l_1 = 50m$,

$$\frac{f_1 l_1}{d_1^5} = 7888000,$$

$f_2 = 0.0425$, $d_2 = 8 \times 10^{-2}$, $l_2 = 500m$,

$$\frac{f_2 l_2}{d_2^5} = 6485000,$$

$f_3 = 0.0442$, $d_3 = 7 \times 10^{-2}$, $l_3 = 200\text{m}$

$$\frac{f_3 l_3}{d_3^5} = 5260000,$$

$$\sum \frac{f_i l_i}{d_i^5} = 19633000,$$

$$\frac{8}{g\pi^2} = 0.08263$$

With these values we make estimates for h and F. The new estimate for Q is calculated using linear regression. The new value of Q can in turn be used to obtain new values of f and $\sum \frac{f_i l_i}{d_i^5}$. These are used to refine new estimators of the root of F. Table T 1.21 below shows how this method progresses to a solution.

T 1.21 Solution of h=25m Darcy Wiesbach

Q	Q^2	h	h-25
0.002	0.000004	6.488852	-18.5111
0.00202	4.08E-06	6.619278	-18.3807
0.004839	2.34E-05	37.97884	12.97884
0.003672	1.35E-05	21.62139	-3.37861
0.003913	1.53E-05	24.55189	-0.44811
0.00395	1.56E-05	25.01641	0.016411
0.003949	1.56E-05	24.99993	-7.4E-05

We can use the Hazen Williams formula to solve the same problem. The Hazen Williams equation for headloss in metric units is given for individual pipes.

Eqn 1.28 $$h = \frac{10.7 l Q^{1.852}}{C_{HW}^{1.852} d^{4.87}}$$

In our configuration the same flow runs through each pipe so the total head is given

Eqn 1.29 $$h = 10.7 Q^{1.852} \sum_{\text{no of pipes}} \frac{l_i}{C_{HWi}^{1.852} d_i^{4.87}}$$

C_{HW} is an arbitrary constant for each pipe that takes into account friction. The value is normally supplied by the manufacturer for new pipes and estimated from experience for old pipes according to their use and operating condition.

Assume we are given a Hazen Williams coefficient of 99 for all three pipes then

$$\sum_{i=1}^{3} \frac{l_i}{C_{HWi}^{1.852} d_i^{4.87}} = 60924$$

and table T 1.22 shows the progression to a solution using initial estimates of 0.002 m³ and 0.00202 m³

T 1.22 Solution of h=25m Hazen Williams

Q	h	h-25
0.002	6.541577	-18.4584
0.00202	6.663243	-18.3368
0.005034	36.15457	11.15457
0.003894	22.47119	-2.52881
0.004105	24.77468	-0.22532
0.004125	25.00554	0.005543
0.004125	24.99999	-1.2E-05

We observe that although the Hazen Williams formula is much easier to use compared to the more analytical Darcy Wiesbach formula, it still gives a relatively close result with about 5% difference.

This is a comparison of solutions obtained using Darcy Weisbach method

$$h = \frac{8 Q_i f_i l_i}{g \pi^2 d_i^5}$$

and Hazen Williams method

$$h = 10.7 \frac{Q^{1.852} l_i}{C_{HWi}^{1.852} d_i^{4.87}}.$$

1.3 Multiple Loop Networks

We are now ready to investigate true elementary networks; the elements of such are nodes that represent a combination of flow supply or demand and junctions between different pipes. The second element of networks is the pipes.

Each node has properties of source or demand (flow into or out of the network). For convention we can set sources negative and demand as positive. Also associated with each node is a list of pipes (at least one) that connect to it.

Each pipe has a diameter(d), length (l), and a coefficient of roughness (e, Darcy Weisbach or C_{HW}, Hazen Williams).

Network calculations are expected to determine a steady state flow in individual pipes and the pressure loss across the length of each pipe due to this flow that will allow equilibrium conditions sustainable over time.

T1.31 Flowchart for analysing Elementary Networks

1. Create a table of nodes, listing whether they are sources or demands or pure connections, list the pipes connected to the respective node.

As the fluid is considered incompressible, the total flow into the node equals the total flow out. With each node there is associated a continuity equation, i.e. the net flow into the node equals the demand/supply. We may treat demands as positive and supplies as negative. If the node represents a reservoir, irrespective of flow direction, source or demand, it may have associated a pressure or free surface water level.

2. List the pipes. Include the diameters, lengths and friction coefficient (e or C_{HW}). Each pipe is given a directional arrow to represent positive flow.

3. We must divide the network into a series of clockwise loops, which in total must not over duplicate, but at the same time must pass through each node and pipe of the network.

This is the first layer towards working towards a solution of our problem. Generally we have n pipes, each with its own flow and headloss this can be represented in a matrix form.

$(F)(Q) = (H)$

F is an (nxn) matrix

Q is a (1xn) column matrix of pipe flows.

H is a (1xn) column matrix generally known.

F is formed from relations of loops and continuity equations.

It may appear that the simple method of solving this type of problem is to create the inverse of F.

$(F^{-1})(F)(Q)=(F^{-1})(H)=(Id)(Q)$

i.e. $Q=(F^{-1})(H)$

However, we have to remember that this is no absolute solution because F is nonlinear and this solution forces approximation of F into a linear function of Q.

An easier solution may be to achieve successive better solutions to the loop equations. These loop equations are independent of the node continuity equations since for each node in the loop, the excess flow into the node would be balanced by an outflow from the node.

A further simplification may be to consider solving just one of the loops at a time. This effectively simplifies the problem to solving the flow of through a succession of pipes, similar to problems solved using Eqn 1.27 and Eqn 1.29. This simplification allows the reduction of a lot of complex computation and accelerates our convergence to a solution.

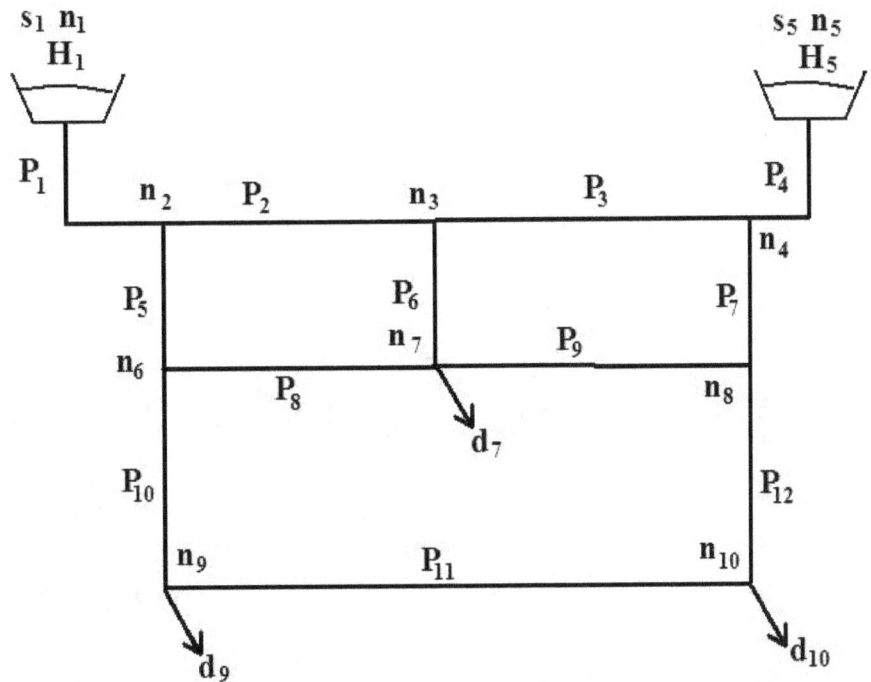

Network example with 10 nodes and 12 pipes

Figure 1.31   Example Network

Lets take an example. Figure 1.31 shows a 12 pipe circuit supplied by reservoirs S1 and S5. We seek a steady state sustainable solution to the network under the conditions given.

D_7= 0.0014m³, D_9= 0.0015m³, D_{10}= 0.0015m³,

H_1 =40m, H_5=5m

We use the flowchart above to analyse the circuit using the steps given in T1.31.

T1.32 List of nodes with properties

node	source/demand	pressure	Pipe connections
1	S1	H1	P1
2	0		P1,P2,P5
3	0		P2,P3,P6
4	0		P3,P4,P7
5	S5	H5	P4
6	0		P5,P8,P10
7	D7		P6,P8,P9
8	0		P7,P9,P12
9	D9		P10,P11
10	D10		P11,P12

The first continuity equation may be to balance total sources and demands.

S1+S5 = D7+D9+D10

written equally,

Q1+Q4 = D7+D9+D10

Nodes 2,3,4,6,7,8,9,10 form the basis of continuity equations, S1 and S5 are supplies that are connected only to single pipes and so are excluded.

T1.33 list the pipes and their properties,

Pipe	diameter	length	e
1	0.06	200	0.0018
2	0.12	2000	0.002
3	0.12	2000	0.0018
4	0.06	250	0.0018
5	0.12	1500	0.002
6	0.06	1500	0.0018
7	0.12	1500	0.002
8	0.06	2000	0.0018
9	0.06	2000	0.0018
10	0.08	1500	0.0018
11	0.08	4000	0.0018
12	0.08	1500	0.0018

Choose positive directions for pipes.

Figure 1.32 below shows how we define the positive directions on pipes of the network.

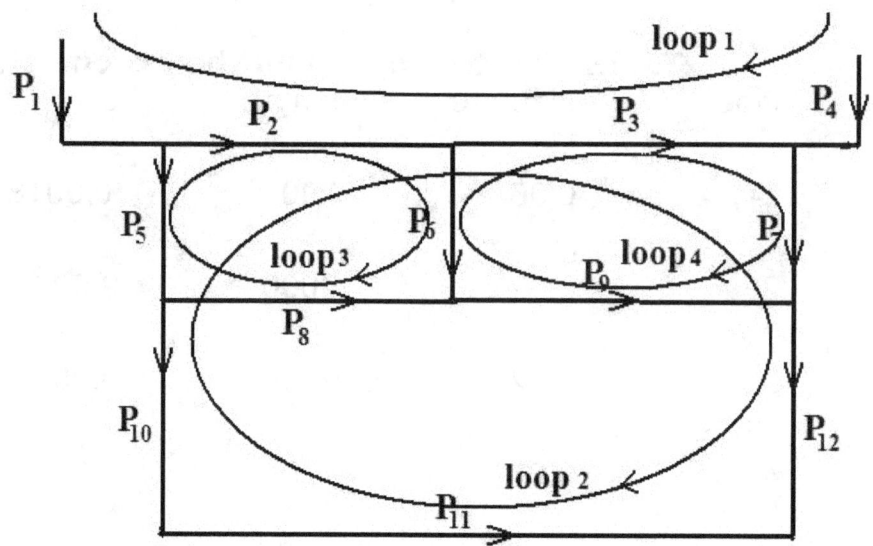

Positive pipe directions & loops for example network

Figure 1.32   Loops and flows

Create all-inclusive loops

T1.34 Network Loops

loop	pipes	headloss
1	-1,-2,-3,4	H5-H1
2	2,3,-5,7,-10,-11,12	0.0
3	2,-5,6,-8	0.0
4	3,-6,7,-9	0.0

In our example we require 4 loops to cover the complete network, note that loop 1 is a "pseudo loop", which is a loop path in the network in which we know the end pressures. In a normal loop, because we end where we start, the pressure difference is 0.0.

With positive pipe directions defined as in figure1.32, we can prepare 8 continuity equations for nodes 2,3,4,6,7,8,9,10. We are now in a situation where we have 8 nodal continuity equations, 1 equation relating total demands and supplies and another 4 loop equations. The 12 pipe flows only require 12 equations to solve them we therefore have a redundancy of one equation. The loop equations are essential in working towards a solution, likewise the demand = supply equation is also fundamental to the solution. One of the 0 demand equations may be ignored with least consequence.

The way we proceed is the process described below.

T1.35 A Darcy Weisbach solution to Pipe Networks
1. We use the continuity equations to create initial flows.
2. Analyse the network into loops with set or zero headloss.
3. Use the flows to calculate pipe coefficients and headloss.
4. Use pipe coefficients and headloss to calculate loop headlosses.
5. Compare the difference between desired headloss and actual headloss for each loop.
6. Select loop with biggest difference.
7. If the difference greater than an acceptable tolerance, then calculate a correcting loop current and add to all pipes in the loop. Go to 3
9. Test results, print results.

The headloss equations are Eqn1.27 for Darcy Wiesbach

The program NsLpG2.exe found in directory www.anglo-african.com/Accessory/NsLpG2 is an implementation of this solution. To work it requires an input file NsLpInG.txt The structure of this input file is such that it has a 5 line header followed by 3 integers that control the units of input, whether metric or imperial, whether diameters are in meters or cms or inches or feet and whether flows are in liters/second or cubic meters/second cubic feet per second or gallons /minute.

Set as" 1 1 0", we are using metric units with diameters in meters and flows in cumecs.

The next line is a single integer that states the number of pipes, np.

The next line lists the pipe number with its characteristics of diameter, length, and roughness e, all in meters. The next line of input is a single integer representing the number of node continuity equations that follow.

Each pipe described has a nominal direction of flow. Flows into the node are tagged +1 flows out of the node are tagged -1. The line ends with the node demand +ive or supply -ve.

Thus node 2 is represented by line

1 -1 0 0 -1 0 0 0 0 0 0 0   0.0

and node 7 with a demand of 0.0044 cumecs is represented by the line

1 0 0 1 0 0 0 0 0 0 0 0 0.0044

Each node equation is represented by a line in the input file. There is an element of symmetry in the node equation array, since for each pipe, according to its assumed direction there is a node which it draws from -1 and a node which it supplies +1. The order of node equations is not important for evaluation. However if there is a redundancy when the loop equations are added we would prefer the loop equation to replace a node equation that has zero demand/supply. For this reason we reorder the node equations with those with zero demand/supply listed last, because these will be first to be replaced by the loop equations.

After the node equations, the following line contains a single integer that represents the number of loop equations. Each loop is represented by a line. Each pipe in the loop is as if it is traversed in a clockwise fashion and is tagged +1 if the pipe is in the direction of the loop or -1 if the pipe direction is counter to the loop. The last value is the expected headloss, which in the case of a closed loop will be 0.00.

Make a copy of NsLpIn.txt and rename it NsLpInG.txt, store it in the same directory as NsLpG2.exe then run NsLpG2.exe. The result is file NsLpOutG.txt.

From the continuity equations the program NsLpG2.exe is able to make an initial estimate for Q the pipe flows, which are subsequently improved by iteration and produces an output file NsLpOut.txt. NsLpG2.exe finds the headloss through loops of pipes and writes the results in NsLpOutG.txt. Which has a header. Below the header output takes the form of pipe no., pipe diameter m, pipe length m, e m, Pipe flow Q m³, bulk fluid velocity v m/s, f, Hl=ΔP/g m, Kp = Hl/Q.

Pipes are input no of pipes, pipe dia M, e M, L M.

T 1.36 Input File Content of NsLpIn.txt

pipe	Dia	L	Q	V	f	Hl	Kp
1	1	0					
12							
1	0.06	200			0.0018		
2	0.12	2000			0.002		
3	0.12	2000			0.0018		
4	0.06	250			0.0018		
5	0.12	1500			0.002		
6	0.06	1500			0.0018		

7	0.12	1500	0.002
8	0.06	2000	0.0018
9	0.06	2000	0.0018
10	0.08	1500	0.0018
11	0.08	4000	0.0018
12	0.08	1500	0.0018

9

0	0	0	0	0	1	0	1	-1	0	0	0	0.0014
0	0	0	0	0	0	1	0	1	0	0	-1	0.0
0	0	0	0	0	0	0	0	0	1	-1	0	0.0015
0	0	0	0	0	0	0	0	0	0	1	1	0.0015
1	0	0	1	0	0	0	0	0	0	0	0	0.0044
1	-1	0	0	-1	0	0	0	0	0	0	0	0.0
0	1	-1	0	0	-1	0	0	0	0	0	0	0.0
0	0	1	1	0	0	-1	0	0	0	0	0	0.0
0	0	0	0	1	0	0	-1	0	-1	0	0	0.0

4

-1	-1	-1	1	0	0	0	0	0	0	0	0	-35

```
0  1  1  0  -1  0  1  0  0  -1  -1  1      0.0

0  1  0  0  -1  1  0  -1 0   0   0  0      0.0

0  0  1  0   0 -1  1  0  -1  0   0  0      0.0

0  0  0  0   0  0  0  1   1  -1  -1 1      0.0
```

The output file lists for each pipe its diameter, length, flow, flow velocity, calculated value f for the pipe flow conditions and the total headloss across the length of pipe.

One can equally use the program Epanet to evaluate the network. This is available as Epanet2 or Epanet3, which is free to download and install from the Epanet web site. It uses a similar process to analyse the network but the results are presented in a different fashion.

The advantage of using the method described is that considering each iteration is only solving a single loop equation, the complexity of calculation is seriously reduced. We only solve the worst fitting loop with the consequence the convergence to a solution is rapid and we are able to use the full Hazen Williams algorithm with a correction for f each time a pipe flow changes.

The benefits of modelling have shown that without raising either a spanner or wrench and without paying for any formal works we are able to anticipate the status of a network in operation. Alterations to the network can immediately be modelled by making changes to the input file. We can test for stress and failure by changing the pipe parameters and node demands. This has obvious advantages in network design and management and is an invaluable tool in water supply analysis.

Chapter 2

A HEURISTIC APPROACH TO SOLUTION OF EQUILIBRIUM IN STATICALLY INDETERMINATE STRUCTURES

This is a heuristic approach to finding the equilibrium flow conditions of a network. It has application in similar problems in non-linear networks and lattices. Any system that can be broken down into distribution of nuclei connected by discrete channels that transfer a property governed by some specific mathematical formulae can be solved using this approach. It is transferable to problems in different disciplines.

1. Solution of non linear Darcy Weisbach equations in a pipe network.

2. Solution of heat conduction and distribution in 3 dimensional lattices.

3. Cell communications and nutrient feed problems in membranes and biological media.

4. Message networks.

5. Transport grids.

6. The distribution of forces and node positions in a lattice structure.

7. The solution of electrical networks.

8. The use of economic models for prediction and analysis of the economy.

This approach is broken down into a number of basic steps.

T2.1 Steps in Forming Approach

1. map out the specific nodes, determine their characteristics and relevant status, such as physical position, whether they are sources or sinks of the property that is in transfer, relative positions, the connections made to each node.
2. Map out the connectors, the properties of the connector that determine its ability to conduct the property in transfer, the nodes it connects.
3. Set up an initial property flow that meets the nodal requirements.
4. Set up a series of loops that traverse each connector at least once.
5. Calculate how the property flows equates with the equilibrium equation for each loop.
6. choose loop with greatest difference from required result.
7. Check if the difference is greater than acceptable tolerance.
8 If greater than acceptable tolerance use mathematical solution to to correct loop and go to 5.
9. Test results, print results.

In the Chapter 1 I have shown how this method has created solutions for pipe networks using Darcy Wiesbach formulation. In this chapter I wish to demonstrate how a modified form of this method can solve statically indeterminate problems in structures. This has importance in engineering when dealing with structures that may be critically deformed by loads, structures such as loaded aerofoils, wings, and hydraulic forms and bridges and possibly towers suffering symmetric or unsymmetrical loads. Structures such that the final loading and shape may be critical to their function or may cause failure.

In such case the nodes are represented by connections between spars and beams. In such a lattice the property transferring between nodes are forces and the channels for transferring the forces are the spars and beams. To ease computation we may consider displacements to be small and approach linearity.

Normally, in a beam we have small displacements described by the figure 2.1 and Eqn 2.1.

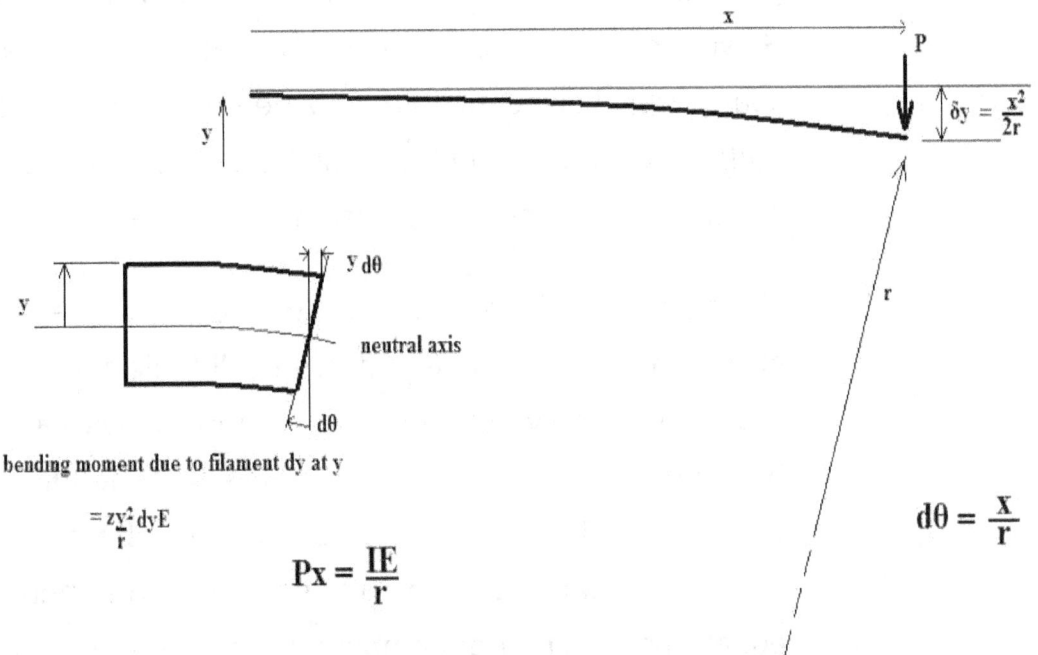

Figure 2.1 Bending Moment

$$\int \frac{Ey^2}{r} z\, dy = Px$$

Eqn 2.1 $\dfrac{EI_y}{r} = Px$

When dealing with elastic beams in compression we can use the formula.

$$F = \frac{\delta l}{l} EA$$

Forces on spars can be resolved into forces tangential and radial to the spar. Radial forces cause the spar to act as a beam and tangential forces cause the spar to act as an elastic member in tension or compression. In such cases the displacements due to beam moments are often much greater than those due to elasticity and usually override. However if the nodes are such that angles between the spars are allowed to vary, no beam bending will happen and forces on the lattice will be solely resolved through elastic tension and compression of the spars.

Since I only wish to demonstrate the efficacy and principle of this method, in my examples I will limit myself to elastic spars that only extend or compress without angular bending moments and the lattice will be limited to 2 dimensions. The same process can easily be extended to 3 dimensions with nodes with fixed beams and bending, with equations with higher complexity and greater computation.

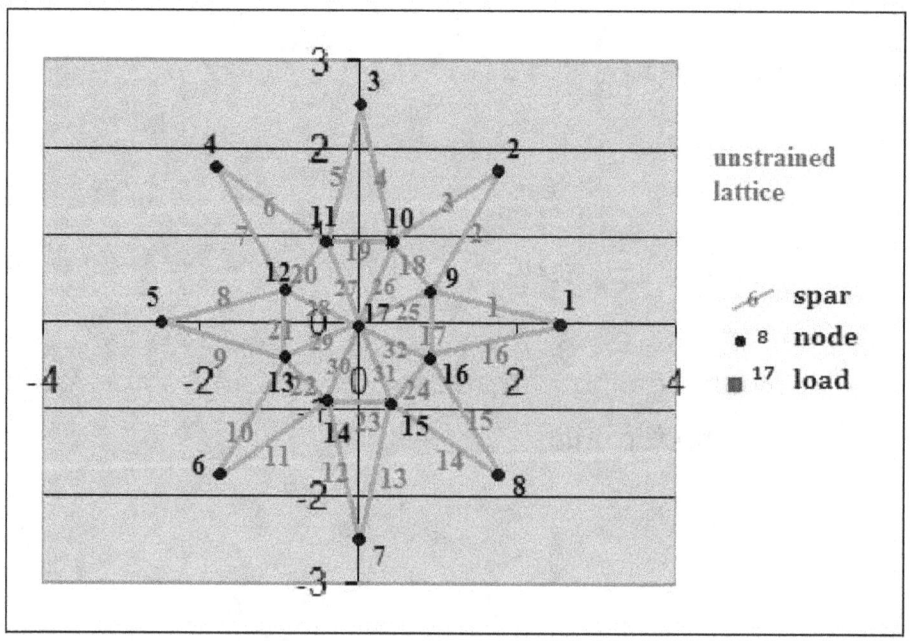

Figure 2.2 Unstrained Lattice

T 2.2 Unstrained Node Positions

node	x	y
1	2.5	0
2	1.768	1.768
3	0	2.5
4	-1.768	1.768
5	-2.5	0
6	-1.768	-1.768
7	0	-2.5
8	1.768	-1.768
9	0.924	0.383
10	0.383	0.924
11	-0.383	0.924
12	-0.924	0.383
13	-0.924	-0.383
14	-0.383	-0.924
15	0.383	-0.924
16	0.924	-0.383
17	0	0

T 2.3 Spar properties

spar	start node	end node	length M
1	1	9	1.622
2	2	9	1.622
3	2	10	1.622
4	3	10	1.622
5	3	11	1.622
6	4	11	1.622
7	4	12	1.622
8	5	12	1.622
9	5	13	1.622
10	6	13	1.622
11	6	14	1.622
12	7	14	1.622
13	7	15	1.622
14	8	15	1.622
15	8	16	1.622

spar	start	end	length M
16	1	16	1.622
17	9	16	0.765
18	9	10	0.765
19	10	11	0.765
20	11	12	0.765
21	12	13	0.765
22	13	14	0.765
23	14	15	0.765
24	15	16	0.765
25	9	17	1
26	10	17	1
27	11	17	1
28	12	17	1
29	13	17	1
30	14	17	1
31	15	17	1
32	16	17	1

Consider the lattice in figure 2.2 that comprises 17 nodes and 32 elastic spars. Nodes 1 to 8 are fixed in position. If node 17 is loaded with a vertical load of 10 N what is the load distribution in each spar and what are the new positions of nodes 9 to 17?

To solve this type of problem I have created the program 2DENet.exe that reads input from the file ENetIn.txt and creates output file EnetOut.txt. 2DENet.exe and supporting files may be found in directory www.anglo-african.com/Accessory/Structs

ENetIn.txt is structured into 2 sections. The first section the number and individual properties of each spar are serially listed. The properties are the id number, the start node, the end node, the E value and the cross sectional area. The second section of the input file details the number and properties of each individual node.

The properties of the node are given, the initial unstressed coordinates of the node, the number of spars attached to the node, the list of spars that attach it, a flag with value 0 or 1 to to indicate whether the node is allowed to reposition freely (0) or if it is anchored into position (1) and finally the x and y coordinates of any loads applied to the node.

For example if we look at spar 18 that connects nodes 9 and 10, the input line for spar 18 is

18 9 10 1.00E+05 1.00E-04

where Young's Modulus is 1×10^5 and cross sectional area is 1×10^{-4}.

Similarly node 17 supports a load of 10 N with start position (0,0) and has 8 spars attaching it to the lattice. The input line for this node is thus.

17   0 0 8 25 26 27 28 29 30 31 32 0 0 -10.0

The whole input file for the conditions given is set out in ENetIn0f10.txt. Copying this to ENetIn.txt and executing 2DENet2.exe, will cause a result file ENetOut.txt to be created. 2DENe.exe uses a modified form of the process set out in earlier in this chapter for solving networks and lattices. It works as set out below.

T 2.4 Program to determine equilibrium positions of free nodes

1. The input file details the unconstrained position of each node when no spar has been extended or compressed by a load, the number of spars adjoining the node and the vector coordinates of loads acting on it are also listed.

2. From the list of spars and their end nodes the program then calculates the unconstrained length of each spar and records it.

3. Set variable representing max unbalanced force in unfixed nodes (MaxF) to 0.0 .

4. For each node the program calculates the net force acting on the node in respect of the extension of each attached spar and the applied load. If the node is flagged as a "non fixed" node the value of resultant force is compared to MaxF. If it has higher magnitude, MaxF is set to this value and the node number is stored as node with highest discrepancy .

5. Having ranged through all the nodes and determined which non fixed node has max resultant force, it is to be repositioned to neutralize the sum of forces acting on it. This is performed by moving the node a calculated amount in the direction of the force so that spars ahead of the node shorten producing a compressive force or reducing the tensile strain. Spars behind the node lengthen either reducing compression or increasing tension. The net result is a force on the node opposing the unbalanced force.

6. check if highest discrepancy is greater than allowed tolerance (max allowed error) if so, go to 3 above.

7. Print results detailing node positions, node status (fixed/non fixed), components and magnitude of resultant force, spars detailing end nodes unstressed length, extension direction, force applied (+ve extension, -ve compression).

File ENetOut23f10.txt shows the results of applying 2DENet.exe to an input file with the content of ENetIn0f10.txt the results are illustrated in figure 2.3 and reviewed below.

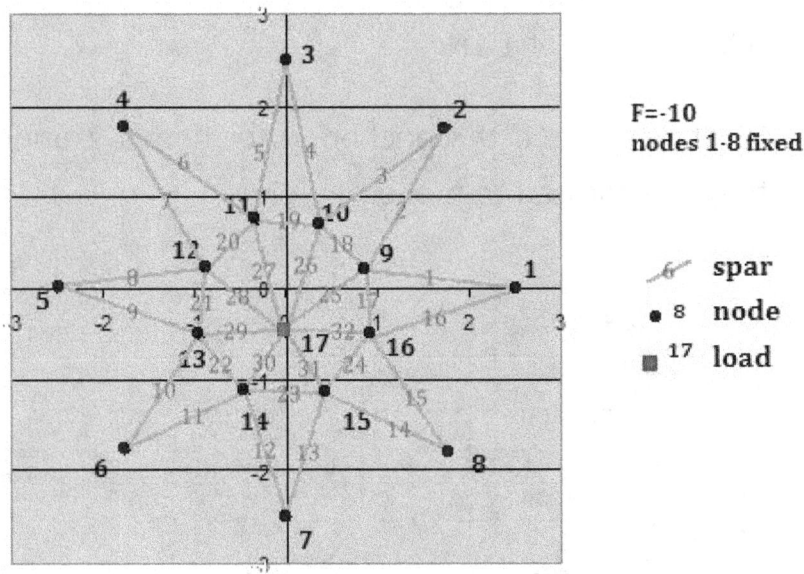

Figure 2.3 Load =-10 N

As expected the main forces are borne on nodes 3 and 7 which combined carry almost half the load, node 3 F=-2.609N, node 7 F=-2.30N. Spars 4,5 and 12,13 share these loads. 4 and 5 in tension with F=1.33N and 12,13 in compression with F=-1.20N. Spars 2,3,6,7 are in extension and 10,11,14,15 in compression bear a substantial part of the load, each carrying of the order of F=1.0N.

It is observed that spars in the upper part of the lattice extend and spars in the lower part of the lattice compress, but there is a load differentiation with the upper spars bearing greater than 10% more load than lower spars. Of the inner ring spars 26,27 and 30,31 bear the greatest loads of up to 2.4N. Spars 25and 28 also bear significant loads of the order of 1.3N. Spars 17 and 21 carry relatively small loads for near vertical spars of about 0.34N and are in compression. Spars 17,18,19,20,21 are in compression;

Spars 22,23,24 are in tension with spar 23 bearing the greatest load of 1.1N.

What happens if the anchoring for nodes 2 and 3 fail and they are allowed to move freely?

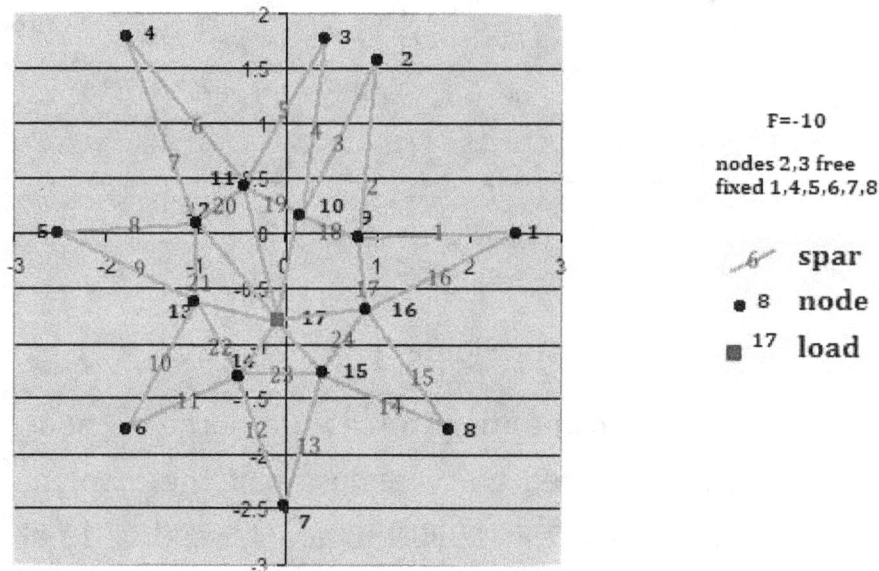

Figure 2.4 nodes 2,3 free

We can see the result of this by changing the fixed/not fixed flag of these nodes to 0 in the input file EnetIn.txt and rerunning the program. the results can be studied in the output file ENetOut23f10.txt and figure 2.4.

A short examination shows, as expected, the symmetry of the lattice is lost. the major part of the load is borne by nodes 4 with 3.2 N and 7 with 3.6N, node 6 bears a substantial load of 2.96N but the greater part of it is in the horizontal direction. Node 8 also bears a substantial load of

1.8N of which 1.58N is vertical and approx -0.87N is horizontal.

Examples given above show an exaggeration of what might be expected with steel or metal spars and members where deformations are limited. What if the materials are of higher elasticity allowing for greater deformations and still behaving in an elastic fashion. One could put in the appropriate values of E and A for each spar that would represent the replaced material, but since this is a study where we require a direct comparison. We may demonstrate such by increasing the load from -10N to -25N.

With the load increased to -25N with nodes 1-8 fixed the results are given in file ENetOut0f.txt and figure 2.5. A short examination of these shows nodes with greatest loading.

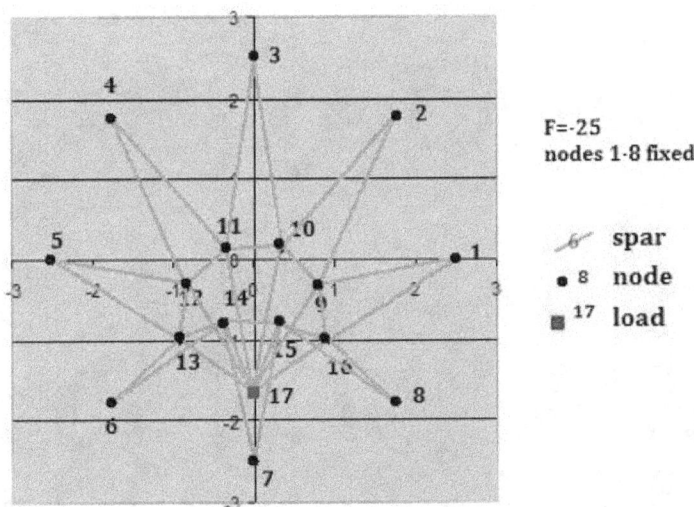

Figure 2.5 force =-25N

T 2.5 nodes with largest forces

node	F
3	9.24 N
2,4	7.42 N
6,8	2.18 N

Nodes 6 and 8 have half their loading in the horizontal direction Fx= 1.52 N.

Let us consider the case of a large load of -25 N and nodes 6,7,8 unsecured and free to reposition as in the case of their anchoring failing. The resultant configuration is shown in the resultant output file ENetOut678f.txt and figure 2.6. We can do a short examination of these files to reveal the salient and relevant points.

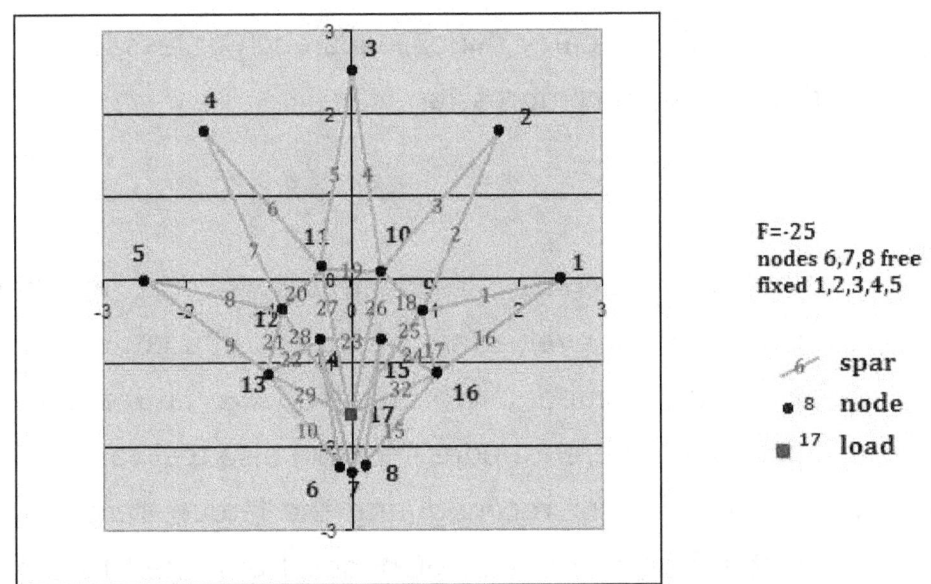

Figure 2.6 Force =-25N nodes 6,7,8 free

As expected the support of node 3 bears the greatest load of -9.52 N nodes 2,4 bear loads of 7.8 N and 5,1 nodes have loads of 2 N. Struts with max stress are detailed below.

T 2.6 Struts with largest forces

Strut	Force
26,27	8.0
2,4,5,7	>4.5 N
3,6	3.35 N
25,28	4.9 N

Struts 17 - 24 carry minor loads each of less than 1 N but they are all in compression.

What this exercise demonstrates is that we have an instrument for evaluating the strains in a lattice connected by elastic members. We can load the lattice with forces applied at different nodes. We can also test for failures. We have options for reconfiguring the lattice, removing spars or changing the status of nodes from fixed to non fixed. This is very useful in design. It details forces throughout the structure, showing the importance of the various members and can be used as an engineering tool for design improvement and testing.

Chapter 3

A LITTLE ROCKET SCIENCE

In this section we will do a bit of speculation on how rockets work mathematically, vis a vis the fuel / payload problem. We need to attack the problem of how much fuel is needed to put a payload into orbit when we know we start with an initial energy position in that the majority of energy is going to be expended accelerating a fuel package which forms the major bulk of the spacecraft.

3.1 Rocket Momentum Consideration

Think on it, a rocket in free space flying with initial velocity. The rocket body and payload flying along with initial velocity V_i. After a small increment of time δt it burns an increment of fuel δm_f. The increase in speed of the bulk body has got to be proportional to the change in momentum of the ejected fuel, proportional to the velocity of ejected gas.

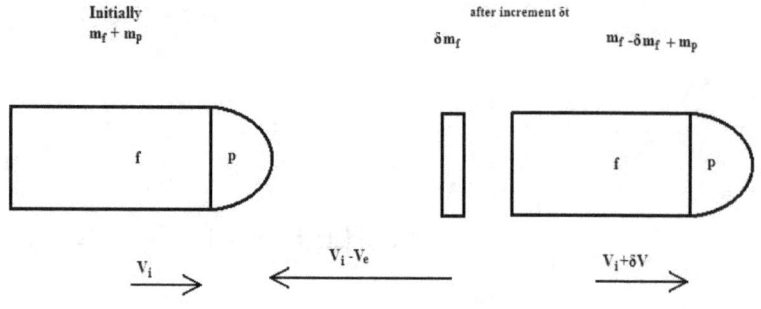

Figure 3.1 Momentum Considerations

momentum

Eqn 3.11 $(m_f + m_p)V_i = \delta m_f (V_i - V_e) + (m_f - \delta m_f + m_p)(V_i + \delta V)$

rearranging

$0 = \delta m_f (V_i - V_e) - \delta m_f (V_i + \delta V) + (m_f + m_p)\delta V$
$\delta m_f (V_e - \delta V) = -(m_f + m_p)\delta V$

since $|V_e| \gg \delta V$ above approximates to

Eqn 3.12 $\delta m_f V_e \approx -(m_f + m_p)\delta V$

Since this happens in increment δt,

Eqn 3.13 $\dfrac{\delta V}{\delta t} \approx -\dfrac{\delta m_f}{\delta t}\dfrac{V_e}{m_f + m_p} = \dfrac{-V_e\left(\dfrac{\delta m_f}{\delta t}\right)}{m_f + m_p}$

This tells us that acceleration of the bulk rocket depends on the ratio of the rate of consumption of fuel to the total rocket mass and the ejection velocity of the fuel.

Writing this as a full differential equation assuming ejection velocity V_e is constant.

$$\int_{V_i}^{V_f} \frac{dV}{dt} dt = -V_e \int_{m_{fi}}^{m_{ff}} \frac{1}{m_f + m_p} dm_f$$

Eqn 3.14 $\quad [V]_{V_i}^{V_f} = -V_e [\ln(m_f + m_p)]_{m_{fi}}^{m_{ff}}$

Evaluating gives

Eqn 3.15 $\quad V_f - V_i = -V_e \ln\left(\dfrac{m_{ff} + m_p}{m_{fi} + m_p}\right)$

This tells us that the difference in velocities at two different times is dependent on the ln of the ratio of the mass of the rocket.

3.2 Energy Conservation

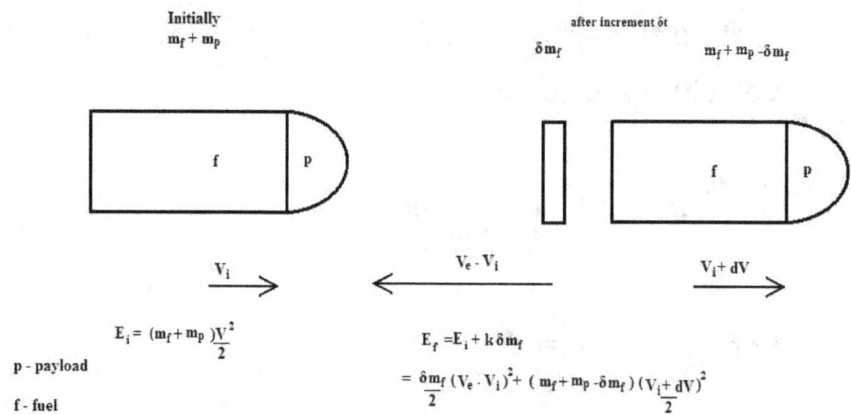

Figure 3.2 Energy Conservation

Eqn 3.21
$$(m_f + m_p)V^2 + 2K\delta m_f = \delta m_f (V_e - V)^2 + (m_f + m_p - \delta m_f)(V + dV)^2$$

Simplifies to

$$2K\delta m_f = (m_f + m_p)(2VdV + dV^2) + \delta m_f (V_e^2 - 2V_eV - 2VdV - dV^2)$$
$$\delta m_f (2K + 2V_eV + 2VdV + dV^2 - V_e^2) = (m_f + m_p)(2VdV + dV^2)$$

since δm_f. and dV are infinitesimal we can drop all terms in the second degree of these quantities.

$$\delta m_f (2K + 2V_eV - V_e^2) = (m_f + m_p)(2VdV)$$

$$\frac{\delta m_f}{m_f + m_p} \approx \frac{VdV}{K + V_e V - \frac{V_e^2}{2}}$$

in integral form

Eqn 3.22 $$\int \frac{dm_f}{m_f + m_p} = \int \frac{VdV}{K + V_e V - \frac{V_e^2}{2}}$$

The left hand integral evaluates

$$\int \frac{dm_f}{m_f + m_p} = {}_{mfi}^{mff}\left[\ln(m_f + m_p)\right]$$

To evaluate the right hand integral we use the identity

$$\frac{V}{K + V_e V - \frac{V_e^2}{2}} = \frac{1}{V_e} - \frac{1}{V_e} \frac{K - \frac{V_e^2}{2}}{K + V_e V - \frac{V_e^2}{2}}$$

$$\int \frac{VdV}{K + V_e V - \frac{V_e^2}{2}} = \frac{1}{V_e} \int \left(1 + \frac{\frac{V_e^2}{2} - K}{K + V_e V - \frac{V_e^2}{2}}\right) dV$$

$$= \frac{1}{V_e}[V] + \frac{1}{V_e} \int \frac{\frac{V_e^2}{2} - K}{K + V_e V - \frac{V_e^2}{2}} dV$$

consider

46

$$\frac{d}{dV}\ln\left(K+V_eV-\frac{V_e^2}{2}\right)$$

$$=\frac{V_e}{K+V_eV-\frac{V_e^2}{2}}$$

so

$$\frac{1}{V_e}\int\frac{\frac{V_e^2}{2}-K}{K+V_eV-\frac{V_e^2}{2}}dV=\left(\frac{V_e^2}{2}-K\right)_{Vi}^{Vf}\left[\ln\left(K+V_eV-\frac{V_e^2}{2}\right)\right]$$

we combine both sides to form

Eqn3.22

$${}_{mfi}^{mff}\left[\ln(m_f+m_p)\right]={}_{Vi}^{Vf}\left[\frac{V}{V_e}\right]+\left(\frac{V_e^2}{2}-K\right)_{Vi}^{Vf}\left[\ln\left(K+V_eV-\frac{V_e^2}{2}\right)\right]$$

The right hand term has two components, If as we may expect in early flight $V_e >> V$ the $\ln(K+V_eV- V_e^2/2)$ term will change very little and the right hand side will be dominated by the V/Ve term.

Thus we can approximate

$${}_{m_{fi}}^{m_{ff}}\left[\ln(m_f+m_p)\right]={}_{V_i}^{V_f}\left[\frac{V}{V_e}\right]$$

This is in effect the same equation as that derived from using to momentum considerations alone (Eqn 3.15), however to make it completely equivalent and practical we have to remember that δm$_f$ is a decrement and we expect V to increase with a decrease in m$_f$, Eqn 3.22 then becomes

Eqn 3.22a
$$-\int \frac{dm_f}{m_f + m_p} = \int \frac{VdV}{K + V_e V - \frac{V_e^2}{2}}$$

with solution

Eqn 3.23
$$-\frac{m_{ff}}{m_{fi}}\left[\ln(m_f + m_p)\right] = \left[\frac{V}{V_e}\right]_{V_i}^{V_f}$$

which is directly equivalent to Eqn 3.15.

Eqn3.15 and Eqn 3.23 show the consumption of fuel is dependent on the velocity ratio V/V$_e$.

V$_e$. is the effective exit velocity of the exhaust gas and is a very important parameter. The rocket should be designed to maximize it within certain constraints. We would expect the exhaust flow to run at about the sonic limit, i.e. the speed of sound, this in turn increases with temperature. The exhaust nozzle of a rocket is important in optimising the exhaust flows, expanding the exhaust gases into a hypersonic cooled state in which they eject into the environment at ambient pressure. It must also have sufficient structural strength to refract the forces expended within it back to the main rocket body. Ambient pressure reduces with altitude; this in turn means that rocket nozzles are optimised to operate within an a designated altitude range.

Fuel types have an associated exit temperature and sonic exit speed. The stoichiometry and chemistry of combustion of a fuel give us a basis for estimating the exhaust velocity, from which we may derive the rocket engine parameter "specific impulse", Isp. This is the impulse delivered per kilogram of fuel. Isp= V_e, the effective exhaust velocity.

Table T3.21 below shows estimates for specific impulse for various fuels.

T3.21 Stoichiometric Isp

fuels	formula	oxidised product	density kg/m^3	Ht per kg fuel+O2 Mj/kg	effective Isp V m/s	Isp/g s^{-1}
Benzene	C6H6		879	10.289	3513.728	358.178
Methane	CH4		0.719	11.124	3653.673	372.444
Hexane	C16H16		660	5.733	2622.839	267.364
Hydrogen	H2		0.09	15.866	4363.353	444.786
Nitrobenzene	C6H5NO2		1175	8.541	3201.419	326.342
P		P2O5		17.316	4558.462	464.675
C		CO2		8.941	3275.573	333.901
Al		Al2O3	2710	16.379	4433.370	451.924
Mg		MgO	1740	14.929	4232.630	431.461
S		SO2		4.635	2358.317	240.399

Specific impulse may also be described in terms of the amount of force delivered per kilogram weight of fuel per second, this gives an indication of the maximum acceleration to be achieved using the fuel. At the Earth's surface the weight of a mass is g x m, g the acceleration due to gravity. Specific impulse when described this way is $\dot{m}V_e/\dot{m}g$, having dimensions of t (sec).

Thus the exit velocity is associated with a change in momentum per unit mass of fuel. The rate of change of momentum of the rocket is $\dot{m}_f V_e$. Rocket motors are designed with considerations to the type of fuel that they burn and the altitude at which they are to operate.

3.3 Energy Considerations

In terms of energy expenditure, the minimum we must expend is the energy required to raise our payload from it's position on the (earth's) surface and into an (earth) orbit with an associated altitude. The brackets indicate that we assume for the sake of example the major body of orbit is the earth, but not necessarily so.

Accurate values of energy of a surface position on the earth depend on geography since the shape of the earth is not absolutely spherical, there may be local contributions due to geological and terrestrial anomalies and any object on it's surface has a diurnal rotation. At a crude level the energy of orbit is the energy associated with the altitude of the orbit and the kinetic energy required to maintain the orbit.

Orbits take the form of "conic sections" and have 3 classes, hyperbolic, parabolic and ellipsoidal. The centre of the major body lies at the focus of the orbit. The trace of the orbit has a point of closest approach, called the perihelion or perigee. The three categories of orbit are according to whether the object orbiting is bound as in an elliptical or circular orbit, or if the object is unbound and only approaches the major body once and then recedes off into the distance as in a hyperbolic or parabolic orbit.

The shape of the orbit is described by the e parameter. This takes a positive value, e less than 1 represents an ellipse. Ellipses are closed repetitive orbits, they have a point of furthest approach called the aphelion or apogee., defined at θ = 0.0 radians. e=0 is the special case of a circular orbit. e=1 represents a parabolic orbit with the internal energy of orbit just sufficient for escape. when e is greater than 1 we produce a hyperbolic orbit.

Our object is to calculate the energy requirements of the orbit in terms of per kilogram of payload.

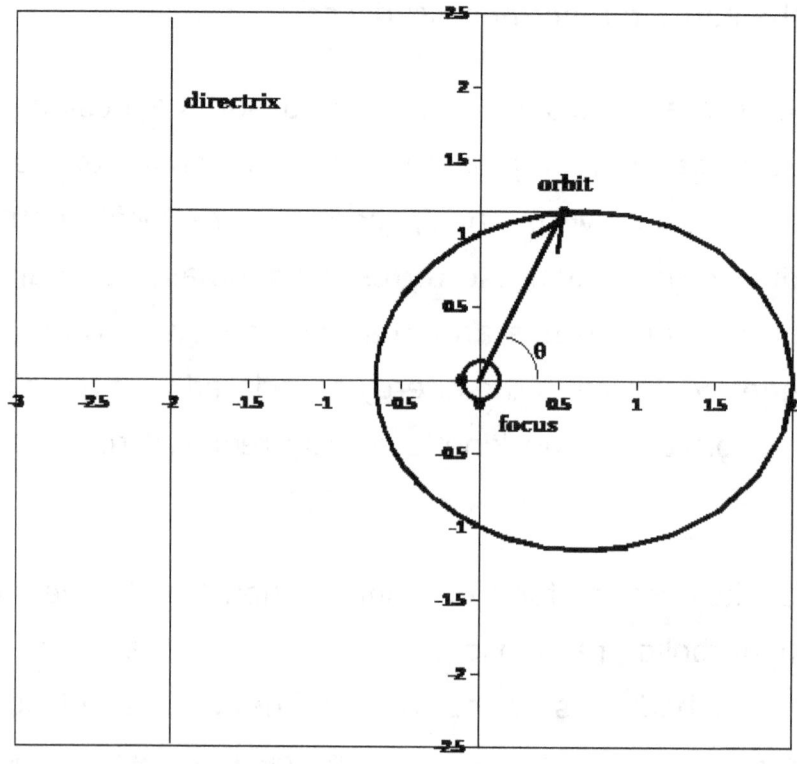

Figure 3.3 Locus of Orbit

the equation of the locus of orbit

Eqn 3.31 $r=\dfrac{l}{1-e\cos\theta}$, l is the semi latus rectum (r=l at θ = π/2)

In polar coordinates the position vector of an object $\vec{r}=r\hat{r}$

With   r – scalar, $\hat{r}$ - unit vector in r direction.

$$r' = \dfrac{d}{dt}\left(r\hat{r}\right)$$

Eqn 3.32 $r' = \dot{r}\hat{r} + r\dot{\theta}\hat{u}$

$$r'' = \ddot{r}\hat{r} + \dot{r}\dot{\theta}\hat{u} + \dot{r}\dot{\theta}\hat{u} + r\ddot{\theta}\hat{u} - r\dot{\theta}^2\hat{r}$$

Eqn 3.33 $r'' = \ddot{r}\hat{r} - r\dot{\theta}^2\hat{r} + 2\dot{r}\dot{\theta}\hat{u} + r\ddot{\theta}\hat{u}$

r'' here has $\hat{u}$ term, but we know the force is central.

Consider $\dfrac{d}{dt}\left(r^2\dot{\theta}\right)$

$$\dfrac{d}{dt}\left(r^2\dot{\theta}\right) = 2r\dot{r}\dot{\theta} + r^2\ddot{\theta}$$

$$2\dot{r}\dot{\theta}\hat{u} + r\ddot{\theta}\hat{u} = \dfrac{1}{r}\dfrac{d}{dt}\left(r^2\dot{\theta}\right)\hat{u} = 0$$

The force is central, only in direction of $\hat{r}$. Thus $h = r^2 \dot{\theta}$ is constant.

Eqn 3.34 $\quad r'' = \left(\ddot{r} - r\dot{\theta}^2 \right) \hat{r} = -\dfrac{Gm}{r^2} \hat{r}$

Now from Eqn 3.31 $\quad r = \dfrac{l}{1 - e\cos\theta}$

$$\dot{r} = \dfrac{-le\sin\theta \, \dot{\theta}}{(1-e\cos\theta)^2} = -\dfrac{h}{l} e\sin\theta$$

$$\ddot{r} = -\dfrac{h}{l} e\cos\theta \, \dot{\theta} = \dfrac{h}{l}\left(1 - \dfrac{l}{r}\right)\dot{\theta}$$

$$= \dfrac{h^2}{r^2 l}\left(1 - \dfrac{l}{r}\right)$$

and

$$r\dot{\theta}^2 = \dfrac{h^2}{r^3}$$

$$r'' = \dfrac{h^2}{r^2 l}\left(\dfrac{l}{r} - 1 - \dfrac{l}{r}\right) = -\dfrac{h^2}{r^2 l} = -\dfrac{Gm}{r^2}$$

Eqn 3.35 $\quad h^2 = Gml = \left(r^2 \dot{\theta} \right)^2$

V can be estimated from Eqn 3.2

$$r' = \dot{r}\hat{r} + r\dot{\theta}\hat{u}$$

$$= \dfrac{h}{l} e\sin\theta \, \hat{r} + \dfrac{h}{r}\hat{u}$$

53

Eqn3.36 $|V|=h\left(\sqrt{\dfrac{e^2\sin^2\theta}{l^2}+\dfrac{1}{r^2}}\right)$

From this we calculate kinetic energy per kilogram of payload

$$\tfrac{1}{2}V^2=\dfrac{h^2}{2l^2}\left(e^2\sin^2\theta+(1-e\cos\theta)^2\right)$$

$$\tfrac{1}{2}V^2=\dfrac{h^2}{2l^2}\left(1+e^2-2e\cos\theta\right)$$

and the total energy per kilogram of payload

$$\int_{r_s}^{r_\theta}\dfrac{Gm}{r^2}dr+\dfrac{h^2}{2l^2}\left(1+e^2-2e\cos\theta\right)$$

$$r_\theta=\dfrac{l}{1-e\cos\theta}$$

$$\Delta E=\left[\dfrac{-Gm}{r}\right]_{r_s}^{\frac{l}{1-e\cos\theta}}+\dfrac{h^2}{2l^2}\left(1+e^2-2e\cos\theta\right)$$

$$\Delta E=Gm\left(\dfrac{e\cos\theta-1}{l}+\dfrac{1}{r_s}\right)+\dfrac{Gm}{2l}\left(1+e^2-2e\cos\theta\right)$$

Eqn 3.37 $\Delta E=Gm\left(\dfrac{e^2-1}{2l}+\dfrac{1}{r_s}\right)$

This shows the difference in energy between resting on the surface and the orbit is a function of m the mass of the body being orbited, e the eccentricity of orbit an l the semi latus rectum.

We may refine the surface energy by the addition of a term for diurnal rotation.

The kinetic energy of rotation of an object per unit mass on the earths surface depends on its latitude.

$$E_{lat} = \tfrac{1}{2}\left(\dot{\Omega} r_s \cos\theta_{lat}\right)^2 = \tfrac{1}{2}\dot{\Omega}^2 r_s^2 \cos^2\theta_{lat}$$

Eqn 3.38 $$E_{lat} = \tfrac{1}{4}\dot{\Omega}^2 r_s^2 \left(1+\cos 2\theta_{lat}\right)$$

The specific surface energy then becomes

Eqn 3.39 $$E_s = \frac{Gm}{r_s} - \tfrac{1}{4}\dot{\Omega}^2 r_s^2 \left(1+\cos 2\theta_{lat}\right)$$

T3.31   Location of Sites

Launch Site	Latitude	Longitude
Baikonur	45.965N	63.305E
Cape Canaveral	28.4556N	80.5278W
New York	40.7142N	74.0064W
London	51.571N	0.1062W
Moscow	55.7517N	37.6008E
Paris	48.8742N	2.3470E

Given $r_s = 6.38 \ast 10^{+6}$ m, $\dot{\Omega} = 7.272 \ast 10^{-5}$

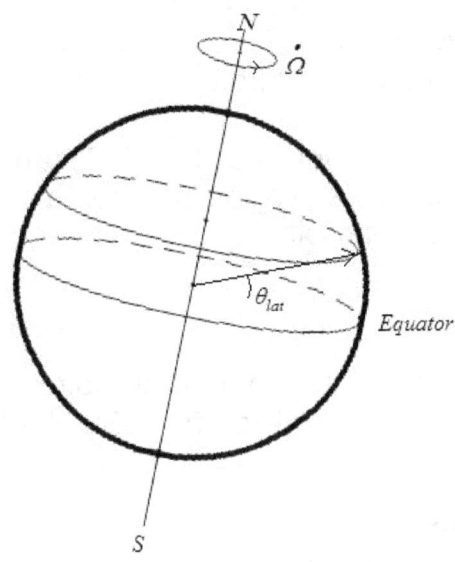

Figure 3.4 Diurnal Rotation

T3.32 Energy of Diurnal Rotation

θ_{lat}	E_{lat} j/kg
90	0.00
80	$3.25*10^{+03}$
70	$1.26*10^{+04}$
60	$2.69*10^{+04}$
50	$4.45*10^{+04}$
40	$6.32*10^{+04}$
35	$7.22*10^{+04}$
30	$8.07*10^{+04}$
25	$8.84*10^{+04}$
20	$9.50*10^{+04}$
15	$1.00*10^{+05}$
10	$1.04*10^{+05}$
5	$1.07*10^{+05}$
0	$1.08*10^{+05}$

It is easily observed that this energy is substantial in latitudes below 40°, but at the same time many of the worlds major cities lie within the latitudes 55° and 40°.

We can now apply our Eqn 3.37 and Eqn3.39 to determine the minimum energies per kg to raise an object from the surface to an orbit.

Ex3.1
Find the energy to launch an object into a low earth circular orbit of 100km altitude from 30° latitude.

r_e=6378 km
G=6.673*10^{-11}
M_e=5.978*10^{+24} kg
l=6.478e*10^{+06} m
e=0.00

Eqn 3.37 $$\Delta E = Gm_e \left(\frac{e^2 - 1}{2l} + \frac{1}{r_e} \right)$$

$$\Delta E = Gm_e \left(\frac{(0-1)}{2*6.478*10^{+06}} + \frac{1}{6.378*10^{+06}} \right)$$

= 3.175*10^{+07} j/kg

We see that the 30° latitude will impart an additional 8*10^{+04} J/kg to the payload. The required energy to put it into orbit is then

(3.175-0.008)* 10^{+07} j/kg

Ex3.2

The orbit the satellite is injected into has a perigee of 100km and an eccentricity of 0.5 all other conditions remain the same as example Ex3.1.

Then l is given from Eqn 3.31 $r = \dfrac{l}{1-e\cos\theta}$

Perigee is given when $\theta = \pi$,

$l = r(1 - e\cos\theta)$

$= 6478*1.5 = 9717$ km

$$\Delta E = Gm_e \left(\dfrac{(0.25-1)}{2*9.717*10^{+06}} + \dfrac{1}{6.378*10^{+06}} \right)$$

$= 4.715*10^{+07}$ j/kg

launched from a 30° latitude will impart an additional $8*10^{+04}$ J/kg to the payload. The required energy to put it into orbit is then

$(4.715 - 0.008)* 10^{+07}$ j/kg $= 4.707*10^{+07}$ j/kg

This orbit has an apogee given when $\theta = 0.0$

$$r = \dfrac{l}{1-e\cos\theta} = \dfrac{9717}{0.5} = 19434 \, km$$

or altitude $(19434 - 6378) = 13056$ km

Eqn 3.37 gives the energy requirement for escape velocity when eccentricity equals 1.

$$\Delta E_{escape} = Gm\left(\frac{e^2-1}{2l} + \frac{1}{r_s}\right) = \frac{Gm}{r_s} = 6.25*10^{+07} \text{ j/kg}$$

T3.33   Escape velocities of celestial bodies

body	Mass kg	Radius m	Gm/r j/kg	V_{escape} m/s
earth	5.98E+24	6.38E+06	6.25E+07	1.12E+04
earth@moon	5.98E+24	3.84E+08	1.04E+06	1.44E+03
sun @earth	1.99E+30	1.50E+11	8.87E+08	4.21E+04
sun @mars	1.99E+30	2.28E+11	5.82E+08	3.41E+04
sun @jupiter	1.99E+30	7.78E+11	1.71E+08	1.85E+04
moon	7.35E+22	1.74E+06	2.82E+06	2.38E+03
mars	6.42E+23	3.38E+06	1.27E+07	5.04E+03

The energy input can be represented by a difference in velocity. Since velocity is relative the energy of a moving object is also relative and converts using the standard relation

$$E = \tfrac{1}{2} mV^2$$

Eqn 3.310 $V = \sqrt{2\dfrac{E}{m}}$

This relation is important in estimating fuel usage because of relations Eqn 3.15 and Eqn 3.23 that relate final velocity to exhaust exit velocity.

In reality the velocity of the bulk rocket is also affected by its position in the gravitational field and the direction of motion, ie the forces of gravity will affect values of measured velocity. By taking energy as the important parameter we change into a scalar property, taking out unnecessary degrees of freedom. Thus energy actually tells us the input required. We may use Eqn 3.310 to relate it to an equivalent velocity that in turn can be V_f the calculated final velocity of our payload. Both velocities in Eqn 3.15 and 3.23 are only representative of energy content and in reality actual measured velocity may be different according to position and relative motion.

T3.34 Effective Specific Impulse

	Energy per kg exhaust	Effective Specific Impulse
Solid Rocket	3.00E+06	1900
Bipropellent Liquid	9.70E+06	3315
Benzene/Oxygen	1.03E+07	3510
Hexane/Oxygen	5.60E+06	2620
H2/Oxygen	1.59E+07	4449

Rocket fuels have an energy of combustion according to their constituents, which in turn controls the maximum exhaust exit velocity and the effective exit velocity. The rocket fuel is considered to be both the fuel and oxidant because both combine in reaction and both are carried on board, as opposed to air breathing engines such as jet engines.

Rocket engines are the most efficient for power/weight ratio because the oxidant is in solid or liquid state, requiring only small pumps for transport and not the bulky and expensive cowls, ducts and compressors of jet engines. An efficient rocket engine works as Carnot engine with about 60% efficiency. The exhaust gas expands adiabatically to up to 40 times its exit volume.

Table T3.35 lists feasible rocket fuels, their effective exhaust velocities and specific impulse. from the table we note the most efficient fuel is hydrogen/oxygen, but we know that liquid hydrogen has its own problems of storage and handling. Bipropellent liquid and benzene/ oxygen are less efficient but the handling and storage of these fuels are much easier to solve. Solid rocket fuel apparently, is least efficient in terms of specific impulse but it has its own advantages, the manufacturing process can be standardized and modularized and the rocket motors are far simpler without moving parts, pumps or ducts.

Returning to examples Ex3.1 and Ex3.2 we are now in a position to work out how much we need to put an object into orbit. From Ex1 we need $3.175*10^{+7}$ j/kg to put a mass into a 100 km low earth orbit from 30° latitude, use Eqn 3.310 to convert this to an equivalent velocity of 7970m/s.

We can then use Eqn 3.15 to work out how much fuel we use.

Similarly for Ex2 we require $4.707 \times 10^{+7}$ j/kg for orbit which is an equivalent velocity of 9700 m/s.

$${}_{m_f+m_p}^{m_p}[\ln m] = -\left[\frac{V}{V_e}\right]_0^{V_f}$$

Eqn 3.311 $$\frac{m_p}{m_f + m_p} = \exp\left(-\frac{V_f}{V_e}\right)$$

T3.35 Volume of fuel for 5000kg Payload

fuel	Effective Specific Impulse	density kg/m³	7970 Ex1 mf	Vol for 5000kg m³	9700 Ex2 mf	Vol for 5000kg m³
Solid Rocket	1900	2209	65.34	177.46	163.89	445.14
Bipropellent Liquid	3315	820	10.07	51.39	17.65	90.10
Benzene/Oxygen	3510	879	8.69	43.28	14.86	74.02
Hexane/Oxygen	2620	659	19.95	98.58	39.54	195.42
H2/Oxygen	4449	67.8	5.00	66.86	7.85	105.00
O2 (LOX)		1141				

With the knowledge of the density of the various fuels And liquid oxygen oxidant we can use the combination of these with estimates for m_f to determine the volume of fuel to place a 5000 kg mass into orbit (representing a 5 tonne payload). This is shown in table T3.36 from this table it is easy to see that bipropellent liquid rocket fuel and benzene have advantage in producing smaller rockets and solid fueled rockets will be largest. This table also shows the advantage of breaking the rocket into different stages, in that the rocket starts off with a bulk of fuel much exceeding the payload. The first stage rocket engines may be optimized for atmospheric pressure and in general would be larger and more powerful than later stages. Stages are jettisoned when they are spent thus unburdening subsequent stages. By breaking the rocket into stages each stage has its motors optimized for their particular task and fueled with the most appropriate fuel. A composite rocket will have stages with various fuels according to use and utility.

This analysis gives a first order estimate of volume and fuel requirement. We have not taken into account allowance for the mass of motors, pumps, storage tanks and auxiliary systems etc. This gives a "ball park" figure that gives a first estimate.

Bibliography

ROGERS GFC & MAYHEW YR (1978) **Engineering Thermodynamics Work and Heat Transfer** 2nd edition London & New York, Longman

JEPPSON R W (1976) **Analysis of Flow in Pipe Networks** Colingwood, Ann Arbor Science

http://www.epa.gov/nrmrl/wswrd/dw/epanet.html

KOTHANDRAMAN CP SUBRAMANYAN S **Heat and Mass Transfer Data Book** (1978) 3rd edition, New Delhi, Bangalore, Bombay, Wiley Eastern Limited

CALVERT JR FARRAR RA (1999) **An Engineering Data Book** Palgrave

YAVORSKY B DETLAF A **A Handbook of Physics** (1980) 3rd edition, Moscow, Mir

www.nasa.gov/centers/wstf/pdf/273192main_rocket_engine_perform_meas4.pdf

SUTTON G P BIBLARZ O (2001) **Rocket Propulsion Elements** 7th edition New York Chichester, Wiley Interscience

PONOMARENKO A (2013) RPA Propulsion-analysis.com/downloads/2/docs/RPA_Assessment of Delivered Performance www.braeunig.us/space/orbmech.htm Basic Space Flight:Orbital mechanics

www.Ins.Cornell.edu/useb/celestia/orbitalparameter.html

en.wikibooks.org/wiki/Astrodynamics/ClassicalOrbit_Elements

en.wikipedia.org/wki/Rocket_engine

www.au.af.mil/primer/rocket_theory.pdf

web.mit.edu/16.unified/www/SPRING/propulsion/notes/node103.html(2008)

REYNOLDS J (1991) **A-Level and AS-Level Mathematics,** London, Longman Group UK limited

www.ingramcontent.com/pod-product-compliance
Lightning Source LLC
Chambersburg PA
CBHW081050170526
45158CB00006B/1919